LARGEST DIGITAL
CAMERA BOOK

The Game Changing Telescope

Shaping Astronomy's Future

SEAN T. ROLAND

COPYRIGHT

TABLE OF CONTENTS

INTRODUCTION

The World's Largest Digital Camera and Its Role in Shaping Astronomy's Future

Humanity's quest to understand the universe has driven centuries of innovation in science, technology, and exploration. From Galileo's pioneering telescope to the Hubble Space Telescope, every new tool has allowed us to peer deeper into space, pushing the boundaries of knowledge and challenging our understanding of the cosmos. Today, this tradition of discovery continues with one of the most exciting advancements in astronomical observation—the Vera C. Rubin Observatory, home to the world's largest digital camera.

Titled *The Game-Changing Telescope Shaping Astronomy's Future*, this book explores how the creation and deployment of this 3.2-gigapixel camera, an unprecedented feat of engineering, is set to revolutionize the field of astronomy. The telescope and its record-breaking digital camera are not just an evolution in technology—they represent a transformative leap in how we observe, analyze, and interpret the universe. This introduction sets the stage for a deeper dive into the mechanics, ambitions, and global impact

4

of the Vera C. Rubin Observatory, with a particular focus on the incredible power of its digital imaging capabilities.

The Evolution of Astronomical Observation

To fully grasp the significance of the world's largest digital camera, we must first consider how far we've come in astronomical observation. The earliest telescopes in the 1600s could only magnify distant objects a few times their original size, allowing astronomers like Galileo to observe Jupiter's moons and Saturn's rings. Over the centuries, technology improved, and we developed ground-based observatories, space telescopes, and, eventually, imaging systems capable of capturing phenomena far beyond the visible spectrum of light.

The advent of digital photography in the 20th century marked a significant turning point in how we observe the universe. For the first time, we could capture vast amounts of data, process it with unprecedented precision, and generate images that surpassed what the human eye could ever perceive. Digital imaging has since become the backbone of modern astronomy, powering projects like the Hubble Space Telescope and the Chandra X-ray Observatory, which have delivered some of the most iconic images of space ever taken.

Now, the Vera C. Rubin Observatory, located in Chile's Atacama Desert—one of the driest and clearest regions in the world—is ready to take digital astronomy to the next level. At the heart of this observatory is the world's largest digital camera, which will capture images with a resolution of 3.2 gigapixels. This chapter introduces the camera's revolutionary design, its critical mission objectives, and its broader significance to science.

The Birth of the Vera C. Rubin Observatory

Named after the pioneering astronomer Vera Rubin, who first provided evidence of dark matter, the observatory is a symbol of her legacy and a harbinger of discoveries yet to come. The observatory's primary mission is to conduct the Legacy Survey of Space and Time (LSST), a comprehensive 10-year survey that will capture the entire southern sky in unparalleled detail. The Rubin Observatory's wide-field, eight-meter telescope, combined with the 3.2-gigapixel camera, will allow astronomers to capture images of the universe with a depth and clarity that has never been achieved before.

The significance of the observatory's location cannot be understated. The Atacama Desert is famous for its clear, dry skies, making it an ideal site for observing faint astronomical

6

objects. The altitude of the observatory, located on Cerro Pachón at 2,682 meters (8,800 feet) above sea level, reduces atmospheric interference, ensuring that images captured by the telescope are crisp and accurate. This strategic positioning is critical, as the observatory aims to capture not just static images but dynamic, time-lapse views of the sky, tracking objects and events as they change over time.

The Largest Digital Camera Ever Built

At the heart of the Rubin Observatory is the LSST Camera, an engineering marvel. Weighing nearly three tons and standing as tall as a small car, this camera is unlike anything the world has seen before. Its 3.2-gigapixel resolution allows it to capture images with unprecedented detail, equivalent to taking pictures with more than 320 million pixels per image. To put this in perspective, the resolution of a typical smartphone camera is around 12 megapixels, meaning that the LSST Camera can capture images that are 266 times more detailed.

But it's not just the sheer number of pixels that makes the LSST Camera special—it's the way the camera can cover vast areas of the sky. With a field of view large enough to capture a region of the sky seven times wider than the full moon in a single shot, the LSST Camera will be able to

7

image large portions of the sky quickly and efficiently. Every few nights, the entire southern sky will be photographed, creating a dynamic movie of the universe in motion.

Each image taken by the camera will be 15 terabytes in size, and the observatory will produce up to 20 terabytes of data every night. Over the course of the LSST's 10-year survey, this will amount to 15 petabytes of data—equivalent to more than 15 million gigabytes. Managing, processing, and analyzing this data is a monumental challenge, but it is one that will pay off in the form of unprecedented discoveries.

The Mission: Mapping the Universe in Real Time

The Legacy Survey of Space and Time is not just about taking beautiful pictures of the night sky. Its primary goal is to study some of the biggest mysteries in cosmology: dark matter and dark energy. Dark matter, which makes up about 27% of the universe's mass, cannot be directly observed, but its presence can be inferred from its gravitational effects on visible matter. Similarly, dark energy, which is thought to drive the accelerating expansion of the universe, accounts for roughly 68% of the universe. Together, dark matter and dark energy make up 95% of the universe, yet we know very little about them.

The LSST Camera will play a key role in unraveling these mysteries. By mapping billions of galaxies and tracking their movements over time, the survey will provide critical data about how mass is distributed throughout the universe and how dark energy influences the expansion of space. The LSST will also detect gravitational lensing—where the light from distant galaxies is bent by the gravity of foreground objects—allowing astronomers to map the distribution of dark matter in unprecedented detail.

But the LSST Camera's mission goes beyond cosmology. It will also monitor near-Earth objects, such as asteroids and comets, providing early warnings for objects that might pose a threat to our planet. In addition, the camera will capture transient astronomical events, such as supernovae, gamma-ray bursts, and the merging of black holes, offering new insights into the most violent and energetic processes in the universe.

A New Era for Big Data in Astronomy

The Vera C. Rubin Observatory's LSST Camera will not only revolutionize astronomy but also herald a new era in how we handle and interpret astronomical data. The sheer volume of data generated by the observatory presents both an opportunity and a challenge for scientists. With 20 terabytes of data being collected every night, traditional methods of data analysis are no longer sufficient. Instead, astronomers will need to rely on cutting-edge technologies like artificial intelligence and machine learning to sift through the data, identify patterns, and make sense of the information.

This massive influx of data will also democratize astronomy in new and exciting ways. Traditionally, access to observatories and their data has been limited to professional astronomers, but the open-access nature of the LSST data will allow anyone, from amateur astronomers to data scientists, to participate in the process of discovery. Citizen science projects will be able to engage the public in analyzing the data, potentially leading to groundbreaking discoveries made by non-professionals.

The Vera C. Rubin Observatory is thus not just a tool for professional astronomers—it is a global resource that will inspire and engage people from all walks of life, encouraging the next generation of scientists, engineers, and explorers.

Setting the Stage for the Future of Space Exploration

As we embark on this new era of astronomical discovery, the Vera C. Rubin Observatory and its world's largest digital camera stand at the forefront of the future of space exploration. By providing unprecedented insights into the universe's most elusive phenomena—dark matter, dark energy, and the life cycles of stars and galaxies—this observatory will not only answer some of the biggest questions in cosmology but will also open the door to entirely new fields of study.

In this book, *The Game-Changing Telescope Shaping Astronomy's Future*, we will delve deep into the mechanics, scientific objectives, and future implications of the LSST Camera. We will explore how this technological marvel will reshape our understanding of the universe, inspire future generations, and usher in a new era of discovery.

The Vera C. Rubin Observatory, with its groundbreaking digital camera, is more than just a technological achievement—it is a symbol of humanity's enduring curiosity and our relentless drive to explore the unknown. As we continue to push the boundaries of what is possible in space exploration, the discoveries made by this observatory will serve as a beacon of knowledge and inspiration for generations to come.

CHAPTER 1

The Dawn of the Rubin Observatory: A Technological Marvel

The Vera C. Rubin Observatory represents one of the most significant leaps forward in the history of observational astronomy. As home to the world's largest digital camera, the observatory is positioned to revolutionize our understanding of the universe through its unprecedented capacity to capture vast sections of the night sky with extreme precision. Located in Chile's Atacama Desert, the observatory stands as a technological marvel, designed not only for scientific discovery but also to overcome the environmental and engineering challenges associated with ground-based astronomy.

This chapter will explore the origins of the Rubin Observatory, its innovative design, and why its location in the Atacama Desert is considered ideal for astronomical observations. We will look at the global collaboration that made this project possible and how the observatory's technology will enable it to tackle some of the biggest mysteries in the universe.

The Vision Behind the Vera C. Rubin Observatory

The Vera C. Rubin Observatory is named after the pioneering astronomer Vera Rubin, who played a crucial role in uncovering the evidence for dark matter. Rubin's work revealed the presence of unseen mass in galaxies, suggesting that dark matter makes up most of the universe's mass. Fittingly, one of the primary objectives of the observatory is to further investigate dark matter and dark energy—two of the most elusive and poorly understood phenomena in cosmology.

The observatory is part of the *Legacy Survey of Space and Time* (LSST), an ambitious 10-year project aimed at mapping the sky and creating the most detailed catalog of celestial objects ever compiled. The LSST project is designed to explore everything from the large-scale structure of the universe to the behavior of asteroids and comets in our own solar system. The Vera C. Rubin Observatory, which houses the LSST Camera, is central to this mission. Over the course of the decade-long survey, the observatory will produce a staggering amount of data, including a time-lapse movie of the night sky that captures how objects and phenomena evolve over time.

The conception of the Rubin Observatory began more than two decades ago, with astronomers and physicists from around the world envisioning a new kind of telescope—one capable of capturing images of the entire sky at unprecedented speed and depth. Early designs for the observatory incorporated technological advances that would allow it to capture extremely wide fields of view and process massive amounts of data, paving the way for the development of the 3.2-gigapixel camera that would become the largest digital camera in the world.

The Construction of the Vera C. Rubin Observatory

Building the Vera C. Rubin Observatory was a colossal undertaking that required years of planning, international cooperation, and overcoming numerous technical and logistical challenges. Construction of the observatory began in 2015, and it quickly became a beacon of innovation in both civil and astronomical engineering.

The observatory itself consists of three main components: the telescope, the camera, and the data processing system. The telescope is a reflecting design with an 8.4-meter primary mirror, making it one of the largest telescopes in the world. This primary mirror works in tandem with a secondary and tertiary mirror to provide a wide field of view

14

while maintaining the sharpness required to capture distant galaxies and faint celestial objects. The telescope's ability to rapidly switch between different regions of the sky allows it to observe dynamic phenomena, such as supernovae, gravitational lensing events, and the movement of near-Earth objects.

One of the most innovative aspects of the observatory's design is its *Simonyi Survey Telescope*. This unique telescope allows the observatory to scan the entire visible sky in just a few nights. The wide-field design enables the telescope to observe a region of the sky about 10 square degrees in size with each image, approximately 40 times the size of the full moon. This feature will enable the observatory to survey the entire southern hemisphere of the sky with extraordinary speed and accuracy.

The observatory's main structure, the dome, is another engineering feat. The dome is designed to be highly efficient in its movement, opening and closing quickly to minimize the time the telescope spends exposed to the elements. The lightweight structure can rotate to follow the movement of the telescope as it tracks objects across the sky, ensuring that observations are conducted smoothly and with minimal interference from external conditions.

The Largest Digital Camera in the World

At the heart of the Rubin Observatory lies the LSST Camera, which is poised to be the most powerful tool ever developed for capturing astronomical images. The camera, which took years to design and build, is capable of producing images with a resolution of 3.2 gigapixels, making it the largest digital camera ever created for astronomical use. To put this into perspective, a single image taken by this camera would be so detailed that it would require 1,500 high-definition TV screens to display it at full resolution.

The camera's construction was a massive technical challenge, requiring advanced optics, sensors, and cooling systems to ensure it could operate under the harsh conditions of the Atacama Desert while maintaining the precision needed to capture images of distant galaxies. The camera consists of 189 individual sensors, each of which captures light in a different part of the electromagnetic spectrum. These sensors work together to create a composite image that captures everything from ultraviolet to infrared light, allowing astronomers to study a wide range of celestial objects in unprecedented detail.

Another remarkable feature of the LSST Camera is its speed. The camera can capture a 15-second exposure of the sky and

16

then be ready to take another image just a few seconds later. This rapid-capture capability is essential for the observatory's mission to create a time-lapse movie of the universe. By taking repeated images of the same regions of the sky over time, the camera will allow astronomers to track how objects and phenomena change, whether it's the explosion of a distant supernova or the shifting positions of asteroids in the solar system.

The data collected by the LSST Camera will be processed and analyzed by a sophisticated data processing system that can handle up to 20 terabytes of data per night. This system will enable astronomers to sift through the massive amounts of information generated by the camera and identify important discoveries in real-time.

The Atacama Desert: A Perfect Location for the Rubin Observatory

One of the key reasons the Vera C. Rubin Observatory is able to achieve such extraordinary scientific goals is its location. Situated on the summit of Cerro Pachón in Chile's Atacama Desert, the observatory is ideally positioned to take advantage of some of the clearest and darkest skies on Earth.

The Atacama Desert, known for its extremely dry conditions and high elevation, provides a near-perfect environment for astronomical observation. The region is largely free from light pollution and has remarkably stable atmospheric conditions, meaning that the turbulence of the air—which can blur images taken by telescopes—is minimized. These qualities allow the observatory to capture exceptionally sharp images of celestial objects, even those that are faint or located at great distances.

The altitude of Cerro Pachón, which rises 2,682 meters (8,800 feet) above sea level, reduces the amount of atmospheric interference between the telescope and the objects it is observing. At this elevation, the observatory is above a significant portion of the Earth's atmosphere, which means that light from distant galaxies reaches the telescope without being scattered or absorbed by the air.

Chile's Atacama Desert is already home to several of the world's most powerful telescopes, including the European Southern Observatory's Very Large Telescope (VLT) and the Atacama Large Millimeter Array (ALMA). These observatories, combined with the Rubin Observatory, make the region a global hub for astronomical research, attracting scientists from around the world to study the universe from this unique vantage point.

The Global Collaboration Behind the Vera C. Rubin Observatory

The Vera C. Rubin Observatory is a product of global collaboration, bringing together scientists, engineers, and institutions from around the world. The observatory is managed by the Association of Universities for Research in Astronomy (AURA) and operated by the National Optical-Infrared Astronomy Research Laboratory (NOIRLab) in the United States. The construction of the observatory was funded by the National Science Foundation (NSF) and the Department of Energy (DOE), with additional support from international partners.

The project's scope and ambition required the expertise of thousands of individuals across multiple disciplines. Astronomers, physicists, engineers, data scientists, and project managers all played a role in bringing the observatory to life. The design and construction of the telescope, mirrors, and camera required cutting-edge technology, while the development of the data processing system involved creating some of the most advanced software and algorithms in existence.

19

The international nature of the Rubin Observatory also extends to its scientific goals. Once the observatory is fully operational, the data it collects will be made available to the global scientific community. This open-access model will allow researchers from around the world to use the observatory's data to conduct their own research, whether they are studying distant galaxies or tracking asteroids in the solar system.

The Rubin Observatory is also committed to public engagement and education. By making its data accessible to the public, the observatory will provide an opportunity for amateur astronomers and citizen scientists to contribute to major discoveries. This approach reflects a broader trend in modern astronomy, where large-scale projects increasingly rely on the participation of non-professional scientists to help analyze data and identify new phenomena.

The Dawn of a New Era in Astronomy

The Vera C. Rubin Observatory marks the beginning of a new chapter in the history of astronomy. Its groundbreaking technology and ambitious mission will transform how we observe and understand the universe, providing insights into some of the most profound questions in science.

From its origins as a visionary idea to its construction in one of the most remote and ideal locations on Earth, the Rubin Observatory stands as a testament to what can be achieved through international collaboration and technological innovation. As it prepares to begin its decade-long survey of the sky, the observatory promises to unlock new The Vera C. Rubin Observatory marks the dawn of a revolutionary era in astronomical observation, leveraging cutting-edge technology and global collaboration to explore the mysteries of the universe in ways never before possible. With its construction in Chile's Atacama Desert, the observatory has already positioned itself as a center piece in the next phase of humanity's exploration of space, equipped with the largest digital camera ever built for scientific purposes. This chapter delves into the rich history and technological marvel behind the observatory, its strategic location in the Atacama Desert, and its transformative potential for the field of astronomy.

A Vision for Revolutionary Astronomy

The origins of the Vera C. Rubin Observatory lie in the global ambition to create a powerful instrument capable of surveying the universe comprehensively. Early 2000s discussions among astronomers highlighted the need for a next-generation observatory that could capture the dynamic universe. The project, initially called the Large Synoptic Survey Telescope (LSST), was rebranded in 2019 to honor Vera Rubin, a pioneering astronomer whose work provided the first observational evidence of dark matter.

Rubin's contributions to the field of astronomy were groundbreaking. Her research on the rotation rates of galaxies revealed discrepancies that could only be explained by the presence of unseen mass, which we now call dark matter. Naming the observatory after her underscores the importance of its mission: exploring the universe's most fundamental mysteries, such as dark matter, dark energy, and the structure of the cosmos.

The Technological Marvel of the Rubin Observatory

At the heart of the Rubin Observatory is the 8.4-meter telescope, designed specifically to capture a wide field of view with exceptional depth. This allows it to image vast sections of the sky quickly, covering areas far larger than most traditional telescopes. However, the true technological marvel lies in its 3.2-gigapixel camera—the largest digital camera in the world—capable of producing images with such high resolution that even the smallest celestial objects can be studied in detail.

The camera features an array of 189 individual sensors, each designed to capture light at different wavelengths, from ultraviolet to infrared. This versatility allows scientists to study a variety of astronomical phenomena, from the movements of asteroids in our solar system to the light emitted by galaxies billions of light-years away. Each image produced by the LSST Camera contains so much detail that it would take over 1,500 high-definition television screens to display a single shot at full resolution.

A key innovation of the Rubin Observatory is its ability to scan the entire sky every few nights, creating a time-lapse "movie" of the universe. This unique capability will allow scientists to track changes in celestial objects over time, such as the explosion of supernovae, the movement of asteroids, and the oscillations of variable stars. This time-domain aspect of astronomy will provide unprecedented insights into the dynamic processes that shape our universe.

The Ideal Location: Chile's Atacama Desert

Choosing the location for the Vera C. Rubin Observatory was no simple task, but ultimately, the observatory found its home in the Atacama Desert, one of the best places on Earth for astronomical observation. Situated on the summit of Cerro Pachón at an elevation of 2,682 meters (8,800 feet), the observatory benefits from some of the clearest, driest, and darkest skies in the world. The desert's unique climate provides ideal conditions for observing faint celestial objects, as it is largely free from light pollution and atmospheric interference.

The Atacama Desert's arid conditions mean that there is minimal water vapor in the atmosphere, which can distort and absorb light as it travels through space. With these optimal conditions, the Rubin Observatory will be able to capture images with incredible clarity, even of distant galaxies and faint cosmic events. Additionally, the high altitude places the observatory above much of the Earth's atmosphere, further reducing the distortions that can affect ground-based telescopic observations.

Chile's Atacama Desert has already established itself as a global hub for astronomy, housing some of the world's most advanced telescopes, including the European Southern Observatory's Very Large Telescope (VLT) and the Atacama Large Millimeter Array (ALMA). The addition of the Vera C. Rubin Observatory to this elite group solidifies the region's importance in the future of astronomical discovery.

Overcoming Construction and Engineering Challenges

Building the Vera C. Rubin Observatory was no small feat. From the harsh environmental conditions of the Atacama Desert to the need for precision engineering, the project faced numerous technical and logistical challenges. Construction began in 2015, and engineers had to overcome everything from the difficulties of transporting large telescope components to the summit of Cerro Pachón to ensuring the structural integrity of the observatory in an earthquake-prone region.

One of the most significant engineering challenges was designing the observatory's dome. Unlike traditional observatory domes that rotate to follow celestial objects, the Rubin Observatory's dome was built to open and close rapidly, minimizing the amount of time the telescope is exposed to the elements. This lightweight structure can move quickly, allowing the telescope to capture fleeting astronomical events, such as gamma-ray bursts, which occur in mere seconds.

The telescope itself was another engineering marvel. Its 8.4-meter primary mirror, cast as a single piece of glass, had to be transported from the United States to Chile—a journey of over 5,000 miles. The mirror was carefully designed to work

25

with the secondary and tertiary mirrors, providing a wide field of view while maintaining the resolution necessary for capturing detailed images of faint objects.

Once completed, the observatory's telescope, dome, and digital camera were integrated into a single system, designed to work in harmony to achieve the observatory's ambitious scientific goals. Every component, from the mirrors to the camera's sensors, was meticulously calibrated to ensure that the observatory would produce images with unprecedented precision.

Global Collaboration and Funding

The Rubin Observatory is a product of international collaboration, bringing together scientists, engineers, and institutions from around the world. The observatory is managed by the Association of Universities for Research in Astronomy (AURA) and operated by the National Optical-Infrared Astronomy Research Laboratory (NOIRLab), based in the United States. Funding for the project comes primarily from the National Science Foundation (NSF) and the Department of Energy (DOE), with additional support from various international partners.

Scientists and engineers from over 30 countries contributed to the design and construction of the observatory. This global collaboration reflects the project's wide-ranging scientific goals, which include not only cosmology and astrophysics but also planetary defense and the study of near-Earth objects. By pooling resources and expertise from around the world, the Rubin Observatory stands as a testament to what can be achieved when the global scientific community works together.

Once operational, the data generated by the Rubin Observatory will be made available to the global scientific community, enabling researchers from all over the world to contribute to new discoveries. This open-access model will democratize astronomy, allowing even amateur astronomers and citizen scientists to participate in the exploration of the universe.

The Scientific Potential of the Vera C. Rubin Observatory

The Rubin Observatory's scientific potential is nothing short of revolutionary. With its ability to scan the entire southern sky in just a few nights, the observatory will provide an unprecedented wealth of data for astronomers. Over the course of its 10-year mission, the observatory will catalog

27

billions of stars, galaxies, and other celestial objects, creating the most detailed map of the universe ever produced.

One of the observatory's primary goals is to study dark matter and dark energy—two of the most mysterious forces in the universe. By mapping the distribution of galaxies and observing how they move, the observatory will provide new insights into how dark matter shapes the structure of the universe. Similarly, by studying the expansion of the universe, the Rubin Observatory will help scientists refine their understanding of dark energy and its role in driving the acceleration of the universe's expansion.

In addition to its cosmological goals, the Rubin Observatory will also contribute to planetary defense. By scanning the sky for near-Earth objects, such as asteroids and comets, the observatory will provide early warnings of potential threats to our planet. Its ability to track the movement of these objects in real-time will be invaluable for developing strategies to mitigate the risk of asteroid impacts.

The Dawn of a New Era in Astronomy

As the Vera C. Rubin Observatory prepares to begin its decade-long survey of the sky, it stands at the forefront of a new era in astronomy. Its groundbreaking technology, combined with its ideal location in Chile's Atacama Desert, will provide scientists with unprecedented insights into the universe's most fundamental mysteries.

The observatory's open-access data model will ensure that the discoveries made by the Rubin Observatory will benefit not just the scientific community, but the world as a whole. Whether it's unlocking the secrets of dark matter and dark energy, tracking near-Earth objects, or simply inspiring the next generation of astronomers, the Vera C. Rubin Observatory is poised to change the way we see the universe.

CHAPTER 2

Breaking Records: The World's Largest Digital Camera

In the annals of astronomical history, the *Legacy Survey of Space and Time (LSST) Camera* stands out as a technological marvel, poised to transform our understanding of the universe. Housed in the Vera C. Rubin Observatory, this 3.2-gigapixel camera is the largest digital camera ever constructed for scientific purposes, with capabilities that will revolutionize how we capture and analyze astronomical data. Its role in the LSST—a decade-long sky survey that will map billions of stars and galaxies—is crucial for investigating some of the universe's deepest mysteries, including dark matter and dark energy.

This chapter will delve into the technical aspects of the LSST Camera, exploring its groundbreaking resolution, construction, and the design features that allow it to capture vast fields of view. We will also look at the advanced sensors, lenses, and cooling systems that make this camera an unparalleled achievement in modern astronomy.

3.2-Gigapixel Resolution: A New Era in Astronomy

The LSST Camera's 3.2-gigapixel resolution is nothing short of revolutionary. To put it in perspective, a single image taken by this camera would be equivalent to 1,500 high-definition TV screens combined. Its resolution allows astronomers to capture vast areas of the sky while still maintaining incredible detail, providing data on objects that are both near and billions of light-years away.

This unprecedented resolution is crucial for the LSST's mission of surveying the sky, as it allows for the detection of faint celestial objects that were previously too dim or distant to be observed. With this resolution, the camera can capture galaxies at the edge of the visible universe, supernovae exploding in distant galaxies, and even asteroids moving within our solar system. Each of these objects can be studied with incredible clarity, enabling new insights into their behavior and properties.

The images produced by the camera are enormous, both in terms of data and scope. Each full-frame image measures 3.5 degrees across, about seven times the width of the full moon as seen from Earth. This wide field of view means the camera can survey the entire southern sky every few nights,

31

building a dynamic, time-lapse movie of the universe that tracks how celestial objects change over time.

The LSST Camera's Construction: Engineering Feats

The construction of the LSST Camera represents one of the most complex engineering feats in modern astronomy. Weighing nearly three tons and standing as tall as a small car, the camera's massive size is necessary to house its intricate system of sensors, lenses, and cooling equipment, all of which must work in perfect harmony to achieve the required performance.

The camera consists of 189 individual sensors, known as charge-coupled devices (CCDs), arranged in a 16.4-inch-wide array. These sensors are capable of detecting light across a wide range of wavelengths, from ultraviolet to infrared. This ability to capture multiple wavelengths in a single exposure is crucial for studying the different physical properties of celestial objects, such as their temperature, chemical composition, and velocity.

Each CCD is responsible for capturing a small portion of the overall image, and when combined, they create a seamless, highly detailed picture of the sky. The quality of the images produced by the LSST Camera depends on the precision with which these CCDs are aligned and calibrated. Engineers have taken extraordinary measures to ensure that the entire system is perfectly aligned, down to the micron level, to avoid distortions and ensure accurate observations.

Another critical component of the LSST Camera is its filter-changing system, which allows it to capture images in six different color bands: u (ultraviolet), g (green), r (red), i (infrared), z (near-infrared), and y (far-infrared). These filters can be rapidly swapped in and out depending on the specific needs of the observation, allowing astronomers to study the same object in multiple wavelengths. The ability to observe celestial objects in different parts of the electromagnetic spectrum is essential for gaining a full understanding of their physical properties.

Capturing Vast Fields of View: Design and Lenses

One of the most striking features of the LSST Camera is its ability to capture vast fields of view without sacrificing image quality. This capability is made possible by an intricate system of three specially designed lenses, each of which plays a critical role in focusing light onto the camera's sensors.

The largest of these lenses is 1.55 meters (5.1 feet) in diameter, making it the largest optical lens ever constructed for an astronomical camera. This primary lens works in tandem with two smaller lenses to focus light precisely onto the array of CCDs. The lens system is designed to minimize optical aberrations—such as blurring or distortion—that can occur when capturing light from distant objects across such a wide field of view.

The precision of these lenses is paramount to the success of the LSST Camera's mission. Each lens was carefully manufactured and polished to exacting specifications, ensuring that they work together to provide sharp, clear images of the sky. Even the slightest imperfection in the lenses could lead to significant distortions in the final images, making the manufacturing process one of the most critical and time-consuming aspects of building the camera.

The wide field of view provided by these lenses allows the LSST Camera to observe large sections of the sky in a single exposure, making it ideal for surveying dynamic events such as supernovae, gamma-ray bursts, and the movement of asteroids and comets. This ability to capture both large-scale and fine details simultaneously is one of the defining features of the camera, making it a powerful tool for studying the universe in unprecedented detail.

Advanced Sensors: The Heart of the LSST Camera

At the heart of the LSST Camera are its advanced sensors, which are responsible for detecting and recording the light from celestial objects. The camera's 189 CCDs are specially designed to capture faint light from distant galaxies while still being sensitive enough to detect relatively close objects, such as asteroids and comets within our solar system.

The CCDs are arranged in a mosaic pattern, allowing the camera to capture a continuous image of the sky. Each CCD is highly sensitive to light and can detect individual photons, the smallest unit of light. This sensitivity is critical for observing faint objects at the edge of the visible universe, where the light reaching Earth is extremely weak after traveling billions of light-years.

The sensors are also designed to operate at very low temperatures to minimize noise and improve image quality. Noise refers to unwanted electrical signals that can interfere with the detection of faint light from celestial objects. By cooling the sensors to temperatures as low as -100 degrees Celsius (-148 degrees Fahrenheit), engineers can reduce noise and ensure that the camera captures the faintest possible signals from the sky.

The cooling system used to maintain these low temperatures is another engineering triumph. The LSST Camera is equipped with a sophisticated refrigeration system that uses liquid nitrogen to keep the sensors cool during observations. This system is designed to operate continuously for the duration of the LSST's 10-year mission, ensuring that the camera can perform at its peak without interruption.

Overcoming Challenges: Precision Engineering and Calibration

Building a camera of this size and complexity required overcoming a number of significant engineering challenges. One of the most difficult tasks was ensuring that the camera's various components—sensors, lenses, filters, and cooling systems—worked together seamlessly to produce high-quality images.

Achieving this required precise calibration of every component, down to the micron level. The alignment of the camera's lenses, in particular, was a major challenge, as even the smallest misalignment could result in distorted images. Engineers used advanced laser alignment techniques to ensure that each lens was perfectly positioned relative to the others, and extensive testing was conducted to verify the accuracy of the system.

The camera's filter-changing system also posed a significant challenge. The system needed to be fast and reliable, capable of swapping filters in and out during the brief intervals between exposures. Engineers designed an automated system that can change filters in less than a minute, allowing the camera to rapidly switch between different wavelengths of light depending on the needs of the observation.

In addition to the mechanical challenges, the LSST Camera required the development of advanced software and algorithms to process the massive amounts of data it generates. Each image captured by the camera is approximately 15 terabytes in size, and the observatory will produce up to 20 terabytes of data per night. Processing this data in real-time is a monumental task, requiring the use of powerful supercomputers and cutting-edge data analysis techniques.

A Game-Changer for Astronomy: The Impact of the LSST Camera

The LSST Camera is not just a technical achievement—it represents a paradigm shift in how we observe and study the universe. Its combination of high resolution, wide field of view, and rapid capture capabilities makes it one of the most powerful astronomical tools ever created.

Over the course of its 10-year mission, the LSST will generate a dynamic map of the sky that will provide unprecedented insights into the behavior of celestial objects and the structure of the universe. The data collected by the camera will be used to study everything from the formation of galaxies to the nature of dark matter and dark energy.

One of the most exciting aspects of the LSST Camera is its ability to capture transient events, such as supernovae and gravitational lensing. These events occur over relatively short timescales and can provide valuable information about the evolution of the universe. By capturing these events in real-time, the LSST Camera will allow astronomers to study them in unprecedented detail.

The LSST Camera will also play a crucial role in planetary defense, as it will be used to track near-Earth objects that could pose a threat to our planet. The camera's ability to rapidly scan large areas of the sky will enable it to detect asteroids and comets that were previously too faint or fast-moving to be observed.

In conclusion, the LSST Camera is a game-changer for astronomy, providing a level of detail, speed, and coverage that has never been achieved before. Itslevel of detail, speed, and coverage that has never been achieved before. Its high resolution, wide field of view, and rapid imaging capabilities will enable it to capture a dynamic map of the universe in unprecedented detail. The technical advancements in its sensors, lenses, and cooling systems represent the pinnacle of modern engineering, pushing the boundaries of what is possible in astronomical observation.

The LSST Camera's role in advancing our understanding of dark matter, dark energy, and the transient universe will be invaluable for cosmology and astrophysics. By capturing data on everything from faint galaxies at the edge of the visible universe to near-Earth asteroids, it is set to be a game-changer in both planetary defense and the study of celestial phenomena.

CHAPTER 3

Unlocking Cosmic Mysteries: Dark Matter and Dark Energy

The universe is a vast, enigmatic place, filled with phenomena that challenge our understanding of physics and the nature of reality. Two of the greatest cosmic mysteries are dark matter and dark energy, which together make up about 95% of the universe's total content, yet remain largely invisible to us. Dark matter, which constitutes approximately 27% of the universe's mass, does not emit, absorb, or reflect light, making it undetectable by conventional means. Dark energy, accounting for about 68%, is even more perplexing, driving the accelerating expansion of the universe. Despite their ubiquity, our knowledge of these components remains limited.

This chapter delves into the scientific objectives of the Vera C. Rubin Observatory's Legacy Survey of Space and Time (LSST) and its 3.2-gigapixel camera, focusing on how these technological marvels will contribute to our understanding of dark matter and dark energy. By capturing detailed images of the sky and utilizing phenomena like gravitational lensing, the LSST will enable scientists to map dark matter

40

and study the effects of dark energy, offering unprecedented insights into the fabric of the universe.

Dark Matter: The Invisible Web of the Universe

Dark matter is one of the most pressing puzzles in astrophysics. Although it cannot be observed directly, its presence is inferred from its gravitational effects on visible matter. Without dark matter, galaxies would fly apart due to their rotational velocities. In fact, the first hints of dark matter came from astronomers like Vera Rubin, after whom the observatory is named, when she observed that the outer regions of galaxies were rotating much faster than expected based on the mass of visible stars and gas alone. Something unseen, some extra mass, was providing the necessary gravitational pull—this was dark matter.

The LSST Camera, with its unparalleled resolution and wide field of view, is uniquely positioned to contribute to our understanding of dark matter. It will capture detailed images of billions of galaxies, allowing astronomers to map the distribution of dark matter by analyzing gravitational lensing effects.

41

Gravitational Lensing: A Window into Dark Matter

Gravitational lensing occurs when light from distant galaxies passes near a massive object, such as a galaxy or a cluster of galaxies, and is bent by the object's gravitational field. This effect creates distorted, magnified, or multiple images of the background galaxy, depending on the mass and shape of the foreground object.

Because dark matter does not emit light, its presence is revealed through gravitational lensing. The amount of distortion in the light from a distant galaxy can tell scientists how much mass is in the foreground object, including both visible matter and dark matter. By mapping the distortions caused by gravitational lensing, astronomers can create a map of dark matter distribution in the universe.

The LSST will observe millions of gravitational lensing events over its 10-year survey. Its ability to capture faint, distant galaxies with extraordinary clarity will provide an unprecedented dataset for studying how dark matter is distributed across the universe. This data will help answer questions such as whether dark matter is smoothly distributed or clumped into structures, how it interacts with visible matter, and whether there are variations in its properties across different regions of the universe.

42

Dark Energy: The Force Behind the Accelerating Universe

While dark matter is responsible for holding galaxies together, dark energy is the mysterious force driving the accelerating expansion of the universe. Discovered in the late 1990s, the existence of dark energy was confirmed when astronomers observed that distant supernovae were dimmer than expected, indicating that the universe's expansion was speeding up rather than slowing down. This discovery was groundbreaking, as it suggested that some form of energy, now termed dark energy, was counteracting gravity on a cosmic scale.

Dark energy is not only difficult to observe, but its nature is completely unknown. Several theories have been proposed to explain it, including the idea that dark energy is related to the cosmological constant—a concept introduced by Einstein—or that it represents a new, unknown form of energy permeating space.

The LSST will play a crucial role in studying dark energy by measuring the expansion rate of the universe over time. By observing distant galaxies and supernovae, the camera will collect data on how fast different parts of the universe are expanding. This will allow astronomers to create detailed

43

models of the universe's expansion history and investigate how dark energy has influenced this expansion over billions of years.

The Role of the LSST in Probing Dark Energy

To study dark energy, the LSST will use several key techniques, including:

1. **Baryon Acoustic Oscillations (BAO):** BAOs are regular, periodic fluctuations in the density of visible matter in the universe, which were created in the early universe due to sound waves moving through the hot plasma. By measuring the distribution of galaxies and how they are separated by these oscillations, scientists can use BAOs as a "standard ruler" to measure the expansion rate of the universe. The LSST will detect BAO signals by mapping the large-scale distribution of galaxies, allowing astronomers to measure how dark energy affects the universe's expansion at different times.

2. **Type Ia Supernovae:** These are a type of exploding star used as "standard candles" because their intrinsic brightness is well understood. By observing how bright these supernovae appear in the sky,

astronomers can calculate their distance and use this information to determine how fast the universe is expanding. The LSST will capture thousands of supernovae over its survey, providing a wealth of data on the expansion of the universe and the influence of dark energy.

3. **Weak Gravitational Lensing:** Just as gravitational lensing can be used to study dark matter, it can also be used to study dark energy. The large-scale structure of the universe is influenced by both dark matter and dark energy, and by observing how galaxies are distorted by gravitational lensing, scientists can measure how dark energy affects the growth of cosmic structures. The LSST's ability to observe weak gravitational lensing in millions of galaxies will provide critical data for understanding the relationship between dark energy and the large-scale structure of the universe.

Mapping the Cosmic Web

One of the key scientific goals of the LSST project is to create a detailed map of the universe's large-scale structure, often referred to as the "cosmic web." This structure consists of galaxies, galaxy clusters, and voids that form a complex

45

network, with dark matter acting as the scaffolding that holds it all together. Understanding how this cosmic web has evolved over time requires precise measurements of both dark matter and dark energy, as they are the dominant forces shaping the universe.

The LSST's wide field of view and ability to capture faint, distant galaxies make it the ideal instrument for mapping the cosmic web. Over its 10-year survey, the LSST will capture images of billions of galaxies across vast distances, allowing astronomers to trace the distribution of both visible and dark matter on a cosmic scale. By studying how galaxies cluster together and how their distribution changes over time, scientists can gain insights into how dark matter and dark energy have shaped the universe's evolution.

Moreover, the LSST will provide data on how dark matter clumps together under the influence of gravity and how dark energy affects the expansion of these structures. This information is essential for testing different theories of dark energy and understanding its impact on the universe's fate.

The Impact of the LSST on Cosmology

The Vera C. Rubin Observatory's LSST Camera represents a major leap forward in cosmology, offering new ways to explore some of the most fundamental questions about the universe. The combination of high-resolution imaging, wide field of view, and time-domain observations will provide astronomers with an unprecedented dataset for studying dark matter, dark energy, and the large-scale structure of the universe.

By mapping the distribution of dark matter through gravitational lensing, the LSST will provide critical insights into how this invisible substance shapes the cosmos. At the same time, its observations of distant galaxies, supernovae, and large-scale structures will shed light on how dark energy influences the expansion of the universe and the growth of cosmic structures.

The data collected by the LSST will also have far-reaching implications for theoretical physics. The precise measurements of dark matter and dark energy provided by the LSST will be used to test existing models of cosmology and potentially reveal new physics beyond the current understanding of the universe. These observations may lead to the discovery of new particles or forces, and they could provide crucial evidence for or against the existence of phenomena like extra dimensions or modifications to general relativity.

A New Era in Cosmic Exploration

The LSST Camera and the Vera C. Rubin Observatory are poised to unlock the secrets of dark matter and dark energy, two of the most elusive and mysterious components of the universe. By capturing detailed images of billions of galaxies and observing the large-scale structure of the universe, the LSST will provide astronomers with the tools they need to map dark matter, study gravitational lensing, and measure the effects of dark energy on the universe's expansion.

As the observatory begins its decade-long survey of the sky, we are on the cusp of a new era in cosmic exploration. The data collected by the LSST will not only deepen our understanding of dark matter and dark energy but also revolutionize the field of cosmology, paving the way for new discoveries that could reshape our understanding of the universe and its fundamental forces.

CHAPTER 4

Tracking the Dynamic Universe: Transients and Variables

The universe is not a static place, but rather a dynamic, ever-changing environment filled with cosmic events that unfold over various timescales. From the sudden explosion of a supernova to the periodic pulsation of a variable star, transient phenomena reveal the underlying processes that govern the cosmos. Observing and understanding these dynamic occurrences is critical to advancing our knowledge of astrophysical processes and the evolution of celestial objects. This chapter will focus on how the *Legacy Survey of Space and Time* (LSST) Camera, installed in the Vera C. Rubin Observatory, will play a transformative role in capturing these transient astronomical events, tracking their evolution, and why monitoring such phenomena is crucial for modern astronomy.

Understanding Transients in Astronomy

In astronomy, transient events refer to phenomena that appear, change, and disappear over relatively short timescales. Unlike steady sources like stars or galaxies that

49

remain relatively constant in brightness and position, transients are temporary in nature, varying from seconds to years. These events include:

- **Supernovae:** Explosive deaths of massive stars.

- **Asteroids and Near-Earth Objects (NEOs):** Small rocky bodies that move through the solar system.

- **Gamma-Ray Bursts (GRBs):** Intense flashes of gamma rays that are often associated with the collapse of massive stars or merging neutron stars.

- **Variable Stars:** Stars that vary in brightness due to internal or external factors, such as binary systems or pulsations.

Each of these events provides invaluable insights into the mechanisms driving the life and death of stars, the evolution of galaxies, and even the future of our solar system. The LSST Camera's unique ability to observe the entire sky frequently and with high resolution makes it an unparalleled tool for tracking and studying these transient phenomena.

The Role of the LSST in Capturing Transient Events

The *LSST Camera* is designed specifically to track transient astronomical events with an unprecedented level of detail.

Its large field of view and fast cadence (the rate at which it takes images of the same part of the sky) allow it to survey the entire southern sky every few nights. This regular monitoring is crucial for capturing dynamic events as they unfold, ensuring that no important transient event goes unnoticed.

Supernovae: Unraveling the Life Cycle of Stars

One of the most important types of transient events that the LSST will track is supernovae. Supernovae are the cataclysmic explosions that occur at the end of a massive star's life. They are key to understanding stellar evolution, the enrichment of the universe with heavy elements, and the overall dynamics of galaxies.

The LSST will be capable of detecting supernovae in distant galaxies as well as those closer to home, providing astronomers with an extensive dataset for studying the different types of supernovae—Type Ia, Type II, and others. Type Ia supernovae, in particular, are important for cosmology because they serve as "standard candles." Their predictable luminosity allows astronomers to calculate their distance and use them to measure the expansion rate of the universe.

In addition to detecting these explosions, the LSST's time-domain observations will allow scientists to follow the evolution of a supernova from its initial brightening to its eventual fading. By tracking the changes in brightness and spectral characteristics, astronomers can infer important details about the progenitor star, the explosion mechanism, and the composition of the ejected material. The LSST's ability to capture large numbers of supernovae over time will revolutionize our understanding of these stellar death events, providing a statistically significant sample that has been unattainable with previous telescopes.

Asteroids and Near-Earth Objects: Monitoring Cosmic Neighbors

Another critical role the LSST will play is in monitoring asteroids and Near-Earth Objects (NEOs). Asteroids are remnants from the early solar system, and studying their composition, trajectories, and interactions provides insight into the conditions of the primordial solar nebula. However, NEOs also represent potential hazards to Earth, as some asteroids and comets could collide with our planet if their orbits bring them too close.

The LSST's wide field of view and frequent sky surveys make it an ideal tool for tracking asteroids and NEOs. The

52

camera will not only be able to detect previously unknown objects but also monitor their positions and movements over time. This will enable astronomers to refine orbital models, predict future trajectories, and assess potential collision risks. The ability to discover and track NEOs with such regularity will be invaluable for planetary defense efforts, allowing for timely warnings and mitigation strategies if an asteroid is found to be on a collision course with Earth.

Moreover, by observing the motion of asteroids within the solar system, the LSST will contribute to our understanding of asteroid dynamics, including how these objects are influenced by gravitational interactions with planets and how they respond to phenomena such as the Yarkovsky effect (the force exerted on an asteroid by the uneven emission of thermal radiation). These observations will provide key insights into the long-term evolution of asteroid orbits and the processes that can alter their paths.

Gamma-Ray Bursts and Other High-Energy Transients

Gamma-Ray Bursts (GRBs) are among the most energetic events in the universe, capable of releasing more energy in a few seconds than the Sun will emit over its entire lifetime. GRBs are thought to be associated with the collapse of massive stars into black holes or the merger of neutron stars.

53

These high-energy transients provide a unique window into extreme astrophysical processes, including the formation of black holes and the production of heavy elements through neutron star mergers.

The LSST's ability to rapidly scan large portions of the sky will allow it to detect the optical afterglows of GRBs and other high-energy transients. When a GRB is detected by space-based gamma-ray observatories like NASA's *Fermi* or *Swift*, the LSST can quickly be pointed in the direction of the burst to capture its optical signature. The time-domain observations provided by the LSST will help astronomers study the evolution of these afterglows and learn more about the environments in which these extreme events occur.

Variable Stars: Decoding Pulsations and Binary Interactions

Variable stars are another class of dynamic objects that will be extensively studied by the LSST. Unlike transients like supernovae that are one-time events, variable stars exhibit periodic or semi-periodic changes in brightness. These variations can be caused by internal processes, such as pulsations, or external factors, such as interactions with a binary companion.

One of the most well-known types of variable stars are Cepheid variables, which play a crucial role in determining cosmic distances. Cepheid variables have a well-established relationship between their pulsation periods and intrinsic luminosities, making them important "standard candles" for measuring distances to nearby galaxies. The LSST will observe thousands of these stars, providing new data on their variability patterns and enhancing our ability to map the structure of the Milky Way and other galaxies.

Another important type of variable star is the eclipsing binary, where two stars orbit each other and periodically block each other's light as seen from Earth. By observing these systems with the LSST, astronomers can gain insights into stellar masses, radii, and evolutionary states. The LSST's continuous monitoring of variable stars will also help identify new classes of variable stars and contribute to our understanding of stellar evolution.

The Importance of Time-Domain Astronomy

Time-domain astronomy—the study of how celestial objects change over time—is becoming increasingly important in modern astrophysics. Traditional astronomical surveys often take snapshots of the sky at a single point in time, but they miss out on capturing the dynamic changes that occur across various timescales. The LSST, with its ability to repeatedly image the entire southern sky every few nights, will provide an unprecedented dataset for time-domain astronomy.

By continuously observing the sky, the LSST will be able to detect short-lived events, such as supernovae and gamma-ray bursts, as well as longer-term changes in variable stars and asteroid motions. These observations will help astronomers track the life cycles of stars, the evolution of galaxies, and the movement of objects within our own solar system.

The LSST's time-domain data will also be crucial for identifying new types of transient events that have never been observed before. As the observatory captures millions of transient events over its decade-long survey, it is likely that astronomers will discover new phenomena that challenge our current understanding of astrophysical processes. By providing a comprehensive view of the dynamic universe, the LSST will open up new avenues of research and lead to groundbreaking discoveries.

Building a Dynamic Map of the Universe

One of the most ambitious goals of the LSST is to create a dynamic, time-lapse map of the universe. Over the course of its 10-year survey, the LSST will capture images of the entire sky every few nights, allowing astronomers to track how objects change and move over time. This will result in a vast dataset that will serve as a living record of the universe's evolution.

This dynamic map will provide valuable information on everything from the formation and evolution of galaxies to the movement of asteroids within the solar system. By studying how celestial objects change over time, astronomers will gain new insights into the processes that drive cosmic evolution and the forces that shape the universe.

The LSST's dynamic map will also be a powerful tool for studying the large-scale structure of the universe. By observing how galaxies cluster together and move over time, the LSST will help astronomers map the distribution of dark matter and study the effects of dark energy on the expansion of the universe. This data will be essential for testing theories of cosmology and understanding the fundamental forces that govern the universe.

A New Frontier in Astronomy

The LSST Camera and the Vera C. Rubin Observatory are poised to revolutionize our understanding of the dynamic universe. By capturing transient events like supernovae, asteroids, gamma-ray bursts, and variable stars The LSST Camera and the Vera C. Rubin Observatory are on the brink of revolutionizing our understanding of the dynamic universe. Through its unprecedented capability to capture transient astronomical events—supernovae, asteroids, gamma-ray bursts, and variable stars—this observatory will provide invaluable insights into the processes that govern the cosmos. As we stand at the dawn of this new era, the observatory's time-domain survey and ability to track the evolution of celestial objects will forever change how we observe the universe.

Tracking Transients: Supernovae and Their Impact on Cosmology

Supernovae are one of the most explosive and illuminating events in the universe. These stellar explosions, which occur at the end of a star's life, provide critical clues about the universe's expansion and the synthesis of heavy elements. Supernovae are categorized into different types based on their progenitors and explosion mechanisms, with Type Ia

58

supernovae playing a particularly important role in cosmology due to their use as standard candles.

The LSST Camera's design allows it to detect supernovae soon after they explode, following their evolution from brightening to fading. By frequently imaging the entire sky, the camera will capture thousands of supernovae over the course of its decade-long survey. Tracking the brightness and spectral evolution of these explosions will help astronomers determine the properties of the progenitor stars, the energy of the explosion, and the composition of the ejected material.

The LSST Camera's ability to observe large numbers of supernovae in diverse environments will provide a more comprehensive understanding of these events. This is crucial for refining models of stellar evolution and improving the accuracy of cosmological measurements. Type Ia supernovae, in particular, serve as vital tools for measuring the universe's expansion rate, helping scientists study dark energy and the accelerating expansion of the universe.

Near-Earth Objects and Planetary Defense

Asteroids and other Near-Earth Objects (NEOs) represent both scientific opportunities and potential threats to Earth. The LSST Camera's ability to track asteroids and map their orbits will contribute to planetary defense efforts by providing early warnings for objects that could pose a collision risk. The frequent imaging and wide field of view of the LSST will allow it to detect previously unknown NEOs and monitor their movements over time.

By tracking the positions and velocities of asteroids, the LSST Camera will help refine models of their orbits, improving our understanding of how these objects move through the solar system. This information will be critical for assessing the likelihood of future impacts and developing strategies to mitigate potential threats. Additionally, studying asteroids offers insights into the composition and history of the early solar system, as these objects are remnants from its formation.

Gamma-Ray Bursts and High-Energy Transients

Gamma-ray bursts (GRBs) are among the most energetic and short-lived events in the universe. These bursts of gamma rays are typically associated with the collapse of massive stars or the merger of neutron stars, and they provide a unique window into extreme astrophysical processes. While space-based observatories like NASA's *Fermi* and *Swift* detect the gamma rays from these bursts, ground-based telescopes like the LSST are essential for observing their optical afterglows.

The LSST's ability to quickly image the sky after a GRB detection will allow it to capture these afterglows and study their evolution over time. By observing the optical counterparts to GRBs, astronomers can learn more about the environments in which these events occur and the physical processes that drive them. The LSST's time-domain observations will provide a wealth of data on these high-energy transients, helping to unlock the mysteries of black hole formation, neutron star mergers, and the production of heavy elements.

Variable Stars: Unraveling Stellar Pulsations

Variable stars are a diverse group of stars that change in brightness due to a variety of physical processes. Some variable stars, like Cepheid variables, pulsate regularly due to changes in their internal structure, while others, like eclipsing binaries, vary in brightness as one star passes in front of another. These periodic changes in brightness provide critical information about the physical properties of the stars, such as their mass, radius, and temperature.

The LSST Camera will be able to monitor variable stars over extended periods, capturing their brightness changes with high precision. This will provide new insights into the behavior of variable stars and their role in stellar evolution. Cepheid variables, in particular, are important for measuring distances in the universe, as their pulsation periods are directly related to their intrinsic brightness. By studying large numbers of Cepheid variables, the LSST will improve our understanding of the distance scale in astronomy and help refine models of galactic structure.

The Power of Time-Domain Astronomy

Time-domain astronomy is the study of how celestial objects change over time. The LSST Camera is uniquely suited for this field of study due to its wide field of view and frequent imaging capabilities. By observing the same regions of the sky every few nights, the LSST will create a time-lapse movie of the universe, capturing transient events and variable phenomena as they unfold.

This ability to track the evolution of astronomical objects is critical for understanding the dynamic processes that shape the universe. Whether it's the explosion of a supernova, the movement of an asteroid, or the pulsation of a variable star, time-domain observations provide valuable data on the physical mechanisms driving these changes. The LSST's time-domain survey will be a game-changer for astronomy, allowing scientists to study the universe in a way that has never been possible before.

Building a Living Map of the Sky

The ultimate goal of the LSST project is to create a dynamic map of the sky that evolves over time. By capturing images of the entire southern sky every few nights, the LSST will provide a continuous record of the universe's changes over its 10-year survey. This living map will be an invaluable resource for astronomers, enabling them to track the motions of celestial objects, identify new transient events, and study the long-term evolution of the cosmos.

This dynamic map will also be a powerful tool for studying the large-scale structure of the universe. By observing how galaxies cluster together and move over time, the LSST will help astronomers map the distribution of dark matter and study the effects of dark energy on the expansion of the universe. These observations will provide critical data for testing cosmological models and understanding the fundamental forces that shape the universe.

The Future of Transient Astronomy

As the LSST prepares to begin its decade-long survey, the field of transient astronomy is on the brink of a major transformation. The LSST Camera's ability to capture transient events with unprecedented detail will open up new avenues of research and lead to groundbreaking discoveries. By providing a comprehensive view of the dynamic universe, the LSST will enable astronomers to study a wide range of transient phenomena, from stellar explosions to the movement of asteroids, and everything in between.

The LSST's data will be made available to the global scientific community, allowing researchers around the world to contribute to the discovery and analysis of transient events. This open-access model will democratize astronomy, providing opportunities for both professional and amateur astronomers to participate in cutting-edge research.

In conclusion, the LSST Camera is poised to revolutionize our understanding of the dynamic universe. By capturing transient events like supernovae, asteroids, gamma-ray bursts, and variable stars, the LSST will provide critical insights into the processes that govern the evolution of celestial objects and the large-scale structure of the universe. As we embark on this new era of transient astronomy, the discoveries made by the LSST will reshape our understanding of the cosmos and unlock new mysteries that have yet to be uncovered.

CHAPTER 5

Building the Universe's Timeline: A 10-Year Survey

The Vera C. Rubin Observatory's Legacy Survey of Space and Time (LSST) represents one of the most ambitious and transformative scientific endeavors in modern astronomy. Spanning a decade, this monumental survey will produce an unprecedented volume of data, capturing the entire southern sky repeatedly with the world's largest digital camera, the 3.2-gigapixel LSST Camera. By systematically mapping billions of celestial objects over ten years, the LSST will create a dynamic, time-lapse view of the universe, allowing astronomers to track changes, discover transient phenomena, and probe the cosmos with a level of detail never before possible.

This chapter will explore the scope of the LSST project, its decade-long survey goals, and the massive datasets it will produce—around 20 terabytes of data per night. We will delve into how this immense trove of information will revolutionize various fields of astronomy, from the study of dark matter and dark energy to planetary defense and the discovery of new celestial objects.

66

The Legacy Survey of Space and Time: A Comprehensive 10-Year Survey

The LSST is designed to capture the most detailed and complete map of the southern sky over the course of ten years, with an emphasis on both depth and breadth. Unlike previous surveys that captured only snapshots of the sky at specific moments, the LSST will repeatedly image the same regions of the sky every few nights, creating a time-lapse record of how celestial objects evolve over time.

The survey will cover an area of about 18,000 square degrees, or roughly half of the night sky. By imaging this vast region repeatedly, the LSST will gather data on billions of stars, galaxies, asteroids, and other celestial objects. The goal is to create a dynamic catalog that not only captures the current positions and brightness of these objects but also tracks their changes over time.

One of the most significant aspects of the LSST is its ability to capture the entire southern sky approximately every three nights. This rapid cadence is critical for detecting transient astronomical events, such as supernovae, gamma-ray bursts, and the movements of asteroids and comets. By continuously monitoring the sky, the LSST will provide real-time data on dynamic events, enabling astronomers to observe and study these phenomena as they unfold.

The Massive Dataset: 20 Terabytes Per Night

One of the most remarkable aspects of the LSST is the sheer volume of data it will generate. Each night, the observatory will capture around 20 terabytes of data—an amount that dwarfs the datasets produced by previous astronomical surveys. Over the course of its 10-year mission, the LSST will produce a total of 15 petabytes of data, equivalent to more than 15 million gigabytes.

This vast amount of data will be stored in a dynamic catalog that can be accessed by astronomers worldwide. The data will include images, object catalogs, and time-series data on the positions, brightness, and spectra of celestial objects. Every night, the LSST will detect millions of new objects, measure the positions and motions of known objects, and identify changes in brightness that could signal transient events like supernovae or variable stars.

The LSST's data processing pipeline is designed to handle this immense dataset efficiently. Using cutting-edge algorithms, artificial intelligence (AI), and machine learning techniques, the system will sift through the data in real time,

68

identifying objects of interest and flagging transient events for immediate follow-up by other observatories. This automated process is critical, as it would be impossible for human astronomers to manually analyze such a large volume of data each night.

Uncovering New Discoveries: The Potential of the LSST Dataset

The massive dataset generated by the LSST will open up new avenues for discovery across multiple fields of astronomy. By creating a time-lapse view of the universe, the survey will allow astronomers to study the evolution of celestial objects in real time, providing valuable insights into the processes that govern the formation and evolution of stars, galaxies, and other cosmic structures.

Probing Dark Matter and Dark Energy

One of the primary scientific objectives of the LSST is to study dark matter and dark energy—two of the most mysterious components of the universe. Dark matter, which makes up about 27% of the universe's mass, cannot be seen directly, but its presence is inferred from its gravitational effects on visible matter. Dark energy, which constitutes about 68% of the universe's total energy, is thought to be responsible for the accelerating expansion of the universe.

The LSST will contribute to the study of dark matter and dark energy by observing the large-scale structure of the universe and the distribution of galaxies. By mapping the positions and movements of billions of galaxies, the LSST will help astronomers trace the distribution of dark matter throughout the cosmos. In addition, the LSST will use gravitational lensing—a phenomenon where light from distant galaxies is bent by the gravity of intervening dark matter—to create detailed maps of dark matter's distribution.

The LSST will also study the effects of dark energy by measuring the expansion rate of the universe over time. By observing distant supernovae and other "standard candles," the LSST will provide new data on how dark energy influences the universe's expansion, helping to refine our understanding of its properties and behavior.

Studying the Formation and Evolution of Galaxies

The LSST's repeated imaging of the sky will allow astronomers to study the formation and evolution of galaxies over cosmic time. By capturing detailed images of galaxies at various stages of their evolution, the LSST will provide new insights into how galaxies form, grow, and interact with one another.

One of the key questions in modern astronomy is how galaxies acquire their mass and evolve over time. The LSST will help answer this question by tracking the growth of galaxies through mergers and accretion of gas. By observing distant galaxies at various redshifts (which correspond to different epochs in the universe's history), the LSST will create a timeline of galaxy evolution, providing new data on the processes that drive their growth.

In addition, the LSST will capture the movement and interaction of galaxies in clusters, providing valuable information on how galaxies interact with one another and with the surrounding dark matter. These observations will shed light on the role of dark matter in shaping the structure of galaxy clusters and the overall large-scale structure of the universe.

Planetary Defense: Tracking Near-Earth Objects

Another critical application of the LSST is in planetary defense. The observatory's ability to rapidly scan the sky will allow it to detect and track Near-Earth Objects (NEOs), such as asteroids and comets, that could pose a threat to our planet. By monitoring the positions and movements of NEOs over time, the LSST will provide early warnings of potential impacts, giving scientists and policymakers the information they need to develop strategies for mitigating the risk.

The LSST's wide field of view and frequent imaging cadence make it an ideal tool for discovering new NEOs and tracking known objects. Over the course of its survey, the LSST is expected to discover hundreds of thousands of new asteroids, including those in potentially hazardous orbits. By monitoring these objects over time, the LSST will provide accurate predictions of their future paths, helping to identify objects that may pose a collision risk.

Discovering Transient Events and Variable Stars

The LSST will also revolutionize the study of transient events—brief, dynamic phenomena that appear suddenly and disappear just as quickly. These events include supernovae, gamma-ray bursts, and variable stars, all of which provide valuable insights into the life cycles of stars and the physical processes that govern the universe.

By repeatedly imaging the same regions of the sky, the LSST will be able to detect transient events in real time, providing astronomers with the opportunity to study them as they unfold. For example, when a supernova is detected, the LSST will capture its brightening, peak, and fading over time, providing a complete record of the explosion. This data will help astronomers learn more about the progenitor stars, the explosion mechanisms, and the aftermath of these cataclysmic events.

In addition, the LSST will provide new data on variable stars—stars that change in brightness over time due to internal or external factors. By observing these stars over extended periods, the LSST will help astronomers study their pulsations, binary interactions, and other variability patterns, providing new insights into stellar evolution.

Revolutionizing Data-Driven Astronomy

The LSST's decade-long survey will produce a dataset unlike anything previously seen in astronomy. The scale of the data—both in terms of volume and complexity—will require new approaches to data analysis and interpretation. Astronomers will need to develop innovative algorithms and computational tools to process and analyze the data in real time, extracting meaningful information from the vast dataset.

Machine learning and artificial intelligence (AI) will play a critical role in this process. These technologies will be used to identify patterns in the data, flag transient events, and classify objects based on their properties. The LSST's data pipeline is designed to process the data as it is collected, providing astronomers with real-time alerts for transient events and other discoveries.

In addition to professional astronomers, the LSST's data will be accessible to the global scientific community, including citizen scientists. The observatory's open-access model will allow anyone with an internet connection to explore the data, potentially leading to new discoveries by non-professionals. Projects like *Zooniverse*, which allow the public to participate in scientific research, will likely play a key role in analyzing the LSST's dataset.

The Promise of New Discoveries

The decade-long LSST survey represents the most comprehensive effort to map the sky ever undertaken. The data collected by the LSST Camera will provide new insights into nearly every aspect of astronomy, from the formation and evolution of galaxies to the behavior of dark matter and dark energy. The observatory's ability to capture transient events in real time, enabling astronomers to study their

74

evolution as they occur, will lead to a new era of discovery in transient astronomy. Whether tracking supernovae, observing variable stars, or monitoring Near-Earth Objects, the LSST will provide unparalleled insight into the dynamic nature of the universe.

As the LSST project begins its decade-long mission, it holds the promise of reshaping our understanding of the cosmos. By creating a detailed timeline of the universe's evolution, it will allow scientists to explore questions that have puzzled humanity for centuries. The survey's contributions to cosmology, planetary defense, and transient astronomy will have far-reaching implications, helping to answer some of the most profound questions in science while raising new ones for future generations to explore.

In conclusion, the LSST's 10-year survey will not only map the universe but also build its timeline—capturing the dynamic processes that shape the cosmos and giving humanity an unprecedented opportunity to understand our place in the vast expanse of space. This undertaking, with its immense data output and potential discoveries, represents a new frontier in data-driven astronomy, ushering in a future filled with groundbreaking discoveries and deeper insights into the workings of the universe.

CHAPTER 6

The Data Revolution in Astronomy: Petabytes of Knowledge

The Vera C. Rubin Observatory's *Legacy Survey of Space and Time* (LSST) is one of the most ambitious astronomical projects ever undertaken. Over the course of its decade-long survey, the observatory will produce a staggering 15 petabytes of data, capturing detailed images of the sky every few nights. This unprecedented volume of data presents a revolutionary opportunity for astronomical research but also introduces significant challenges in data storage, processing, and analysis. The sheer scale of this dataset, equivalent to over 15 million gigabytes, has ushered in a new era of data-driven astronomy, requiring cutting-edge computational techniques, artificial intelligence (AI), and machine learning (ML) to handle and interpret the information effectively.

In this chapter, we will explore the data revolution that the LSST will bring to astronomy, focusing on the computational challenges of processing such a massive trove of information. We will also discuss how AI, machine learning, and other advanced technologies will play an essential role in analyzing the data, enabling astronomers to make groundbreaking discoveries.

The Scale of the LSST Dataset: 15 Petabytes of Information

To fully appreciate the magnitude of the LSST's data output, it's important to understand what 15 petabytes represent. A single petabyte is equal to 1,024 terabytes or over one million gigabytes. The LSST will generate 20 terabytes of data each night, collecting more data in a week than many previous astronomical surveys did in their entire lifetimes. Over ten years, this adds up to 15 petabytes of data—enough to fill over 200,000 Blu-ray discs.

The LSST's data will include high-resolution images of billions of celestial objects, spanning a vast range of distances and scales. From nearby asteroids to galaxies billions of light-years away, the LSST will provide an unprecedented level of detail for every object it observes. This massive dataset will be used to study everything from the structure of the Milky Way to the nature of dark matter and dark energy, but the sheer volume of information presents a unique set of challenges.

Data Processing Challenges

Processing 15 petabytes of data presents significant computational challenges. The raw data collected by the LSST must be processed into a form that astronomers can use for scientific analysis, including extracting meaningful information about celestial objects and identifying transient events such as supernovae and asteroids. This process involves several steps, including image calibration, object detection, and classification.

1. **Image Calibration and Cleaning**: Raw astronomical images contain various forms of noise and distortion, which must be corrected before the data can be used. This includes compensating for atmospheric turbulence, telescope imperfections, and sensor noise. The LSST's data processing pipeline must perform these corrections automatically for each of the millions of images captured by the telescope.

2. **Object Detection and Classification**: Once the images are calibrated, the next step is to detect and classify the objects in each image. The LSST will capture billions of stars, galaxies, and other celestial objects, each of which must be identified and

78

cataloged. This requires sophisticated algorithms capable of distinguishing between different types of objects and separating real objects from artifacts or noise.

3. **Time-Domain Analysis**: A key feature of the LSST is its ability to capture transient events—objects that change over time, such as supernovae, asteroids, and variable stars. The LSST will repeatedly image the same regions of the sky every few nights, creating a time-lapse record of how these objects evolve. The data processing pipeline must be able to track these changes and identify transient events in real time.

Given the volume of data and the complexity of these tasks, traditional data processing techniques are insufficient. Instead, the LSST must rely on state-of-the-art computational infrastructure, including powerful supercomputers and distributed computing systems, to handle the data. The observatory's data management system is designed to process the data in real time, ensuring that astronomers can quickly access the latest observations and detect transient events as they occur.

AI and Machine Learning in Astronomical Data Analysis

One of the most significant advancements in the field of data-driven astronomy is the application of artificial intelligence (AI) and machine learning (ML) techniques to analyze large datasets. The LSST's 15-petabyte dataset is too vast for human astronomers to analyze manually, making AI and ML essential tools for extracting meaningful information from the data.

Machine Learning for Object Classification

One of the primary uses of machine learning in the LSST project is object classification. The LSST will capture images of billions of objects, ranging from stars and galaxies to asteroids and comets. Manually classifying these objects would be impossible, given the volume of data, but machine learning algorithms can be trained to recognize different types of objects based on their appearance in the images.

By using large training datasets of labeled objects, machine learning models can learn to distinguish between different types of celestial objects, even when the images are noisy or incomplete. These models can then be applied to the LSST's data in real time, automatically classifying objects as they are detected and adding them to the survey's catalog.

Machine learning also allows astronomers to identify rare or unusual objects that may have been missed by traditional methods.

AI for Detecting Transient Events

Another critical application of AI in the LSST project is the detection of transient events. Transients are astronomical objects or phenomena that appear and change rapidly, such as supernovae, gamma-ray bursts, and asteroids. Detecting these events in real time is one of the primary goals of the LSST, but the sheer number of images captured each night makes manual detection impractical.

AI algorithms can be trained to recognize the signatures of transient events in astronomical images, allowing them to detect and flag these events as they occur. These algorithms can process the LSST's data in real time, providing alerts to astronomers when a new transient event is detected. This enables rapid follow-up observations by other telescopes, ensuring that these fleeting events are not missed.

One of the challenges in detecting transients is distinguishing between real events and false positives, which can be caused by noise, sensor errors, or other artifacts in the images. Machine learning algorithms are particularly well-

81

suited to this task, as they can be trained to recognize the differences between real events and false positives, improving the accuracy of transient detection.

Deep Learning for Galaxy Evolution and Cosmology

Deep learning, a subset of machine learning, has become an increasingly important tool in the analysis of astronomical data. Deep learning algorithms, particularly convolutional neural networks (CNNs), have proven highly effective at recognizing complex patterns in images, making them well-suited to tasks such as galaxy classification and the study of galaxy evolution.

The LSST's decade-long survey will capture detailed images of billions of galaxies, providing an unparalleled dataset for studying how galaxies form, grow, and evolve over time. Deep learning models can be used to analyze these images, identifying patterns in galaxy morphology, structure, and behavior that may provide insights into the processes that drive galaxy evolution.

In addition, deep learning techniques are being applied to cosmological research, including the study of dark matter and dark energy. By analyzing the large-scale structure of the universe and the distribution of galaxies, deep learning models can help astronomers better understand the role of dark matter in shaping cosmic structures and the effects of dark energy on the expansion of the universe.

Distributed Computing and Cloud Infrastructure

The LSST's data processing challenges are not limited to the complexity of the analysis. The sheer volume of data also requires massive computational resources for storage and processing. To meet these demands, the LSST project has developed a distributed computing infrastructure that leverages cloud-based technologies and supercomputers to handle the data.

1. **Supercomputing Clusters**: The LSST will rely on powerful supercomputing clusters to process the raw data generated each night. These clusters are designed to handle the massive computational workload of image calibration, object detection, and time-domain analysis. By distributing the workload across multiple nodes, these supercomputing systems can process the data in parallel, significantly reducing the time required for data analysis.

2. **Cloud Computing**: In addition to supercomputers, the LSST project is exploring the use of cloud

computing platforms to store and process its data. Cloud-based infrastructure provides the scalability needed to handle the growing dataset, allowing the LSST to dynamically allocate resources based on demand. This is particularly important for processing large volumes of data in real time, such as when a transient event is detected.

3. **Data Archiving and Access**: Storing 15 petabytes of data requires an efficient and secure archiving system. The LSST project has developed a distributed data archive that will store the entire dataset and make it accessible to astronomers around the world. This open-access model ensures that the LSST's data can be used by researchers from different institutions, fostering collaboration and enabling new discoveries.

The Impact of the LSST on Data-Driven Astronomy

The LSST represents a paradigm shift in the field of astronomy, where data-driven research will play an increasingly central role. The massive dataset generated by the LSST will enable astronomers to tackle questions that were previously out of reach, from understanding the nature of dark matter and dark energy to discovering new

84

exoplanets and tracking asteroids that could pose a threat to Earth.

The LSST's data revolution will also have a profound impact on how science is conducted. Rather than focusing on small, targeted datasets, astronomers will have access to a vast, comprehensive survey of the sky, allowing them to study a wide range of phenomena simultaneously. The combination of big data, AI, and machine learning will enable new discoveries at a scale and speed that were unimaginable just a few decades ago.

In addition, the LSST's open-access model will democratize astronomy, making the data available to researchers, students, and citizen scientists around the world. This will foster a new era of collaboration, where scientific breakthroughs across institutions will be accelerated. The LSST's data will serve as a resource not just for professional astronomers but for citizen scientists and enthusiasts eager to contribute to significant discoveries.

The Future of Data-Driven Astronomy

The Vera C. Rubin Observatory's Legacy Survey of Space and Time represents a milestone in the intersection of astronomy and data science. Over the next decade, the observatory will generate an unprecedented 15 petabytes of data, capturing the evolution of the universe in ways never before possible. The scale of this endeavor requires cutting-edge computational solutions, from advanced machine learning algorithms to distributed cloud computing infrastructure, to manage, process, and analyze this vast dataset.

AI and machine learning will play an increasingly vital role in making sense of this data, helping astronomers classify objects, detect transient events, and uncover hidden patterns in the cosmos. The integration of these technologies will enable real-time analysis, ensuring that key discoveries are identified and shared with the scientific community as they happen.

As we move further into the era of data-driven astronomy, the LSST will be a key player in revolutionizing our understanding of the universe. By embracing the data revolution, astronomers will unlock new insights into the nature of dark matter, dark energy, the formation of galaxies, and the dynamic events that shape the cosmos. The LSST's contributions to the field will resonate for generations to come, marking a new chapter in humanity's quest to explore and understand the universe.

CHAPTER 7

Impact on Near-Earth Object Detection: Securing Our Solar System

The Vera C. Rubin Observatory's *Legacy Survey of Space and Time* (LSST) holds immense potential for planetary defense, particularly in the detection and tracking of Near-Earth Objects (NEOs). These NEOs—asteroids and comets whose orbits bring them close to Earth—pose a potential risk to our planet, and understanding their trajectories is critical for assessing impact hazards. Over its 10-year survey, the LSST will provide an unparalleled ability to monitor these objects, enabling the early detection of potential threats and contributing to the security of our solar system.

This chapter will explore how the LSST Camera's capabilities will enhance NEO detection, discussing the role it plays in monitoring small solar system bodies and how it contributes to global planetary defense initiatives. We will also delve into how this project fits into the broader efforts of international space agencies and organizations tasked with protecting Earth from celestial threats.

The Threat of Near-Earth Objects: An Overview

Near-Earth Objects (NEOs) include asteroids and comets that pass within 1.3 astronomical units (AU) of Earth, with 1 AU being the distance from Earth to the Sun (approximately 93 million miles). While most NEOs are relatively small and pose no immediate threat, some have orbits that could potentially lead to collisions with Earth. An impact from a large NEO could cause catastrophic damage, similar to the event that led to the extinction of the dinosaurs 66 million years ago.

The scientific community has long recognized the need to monitor these objects and track their orbits to identify any potential risks. The discovery of NEOs and monitoring of their trajectories has traditionally been done using a combination of ground-based telescopes and space missions. However, many smaller objects remain undetected due to limitations in observational coverage and technology.

The LSST is poised to address these challenges by providing a systematic and comprehensive survey of the sky. Its powerful camera, coupled with its frequent imaging cadence, will allow it to detect and track thousands of NEOs with unprecedented accuracy. This will significantly

88

enhance our ability to identify potential threats early, allowing for timely responses and risk mitigation.

The Role of the LSST in NEO Detection

One of the key scientific goals of the LSST is to contribute to planetary defense by detecting and tracking NEOs. The observatory's ability to image large portions of the sky every few nights makes it an ideal tool for discovering these small, fast-moving objects. The LSST Camera's wide field of view—roughly 3.5 degrees across—allows it to cover vast areas of the sky in a single exposure, while its high-resolution sensors provide the detail needed to accurately track the positions of objects over time.

The LSST's repeated imaging of the same regions of the sky will enable it to detect changes in the positions of NEOs, helping astronomers calculate their orbits and predict future movements. This is particularly important for identifying objects that could pose a threat to Earth, as even small deviations in an NEO's trajectory can lead to significant changes in its path over time.

In addition to detecting new NEOs, the LSST will also play a critical role in tracking known objects. By monitoring their positions over time, the observatory will provide updated orbital data, allowing scientists to refine their predictions and assess any potential risks more accurately. The LSST's ability to observe faint objects will also enable it to detect smaller NEOs that have been missed by previous surveys, filling in gaps in our knowledge of the NEO population.

Discovering New NEOs: Expanding Our Catalog

The discovery of new NEOs is one of the most important contributions the LSST will make to planetary defense. Estimates suggest that there are more than 25,000 NEOs larger than 140 meters in diameter, of which only a fraction have been discovered. These larger objects are of particular concern, as an impact from one of them could cause widespread devastation.

However, even smaller NEOs can pose a significant threat, as demonstrated by the 2013 Chelyabinsk meteor event in Russia. In that incident, a 20-meter-wide asteroid exploded in the atmosphere, releasing energy equivalent to approximately 30 times that of the atomic bomb dropped on Hiroshima. The explosion caused widespread damage, injuring over 1,500 people.

The LSST will help expand the catalog of known NEOs by detecting objects down to smaller sizes, providing critical

90

information about the population of potentially hazardous objects. Its ability to survey the entire southern sky every few nights ensures that no region of space is left unobserved for long, increasing the chances of detecting NEOs that might otherwise go unnoticed.

Moreover, the LSST's sensitivity to faint objects will allow it to detect NEOs that are farther from Earth or have low surface brightness. This will provide a more complete picture of the NEO population, helping scientists better understand the distribution and characteristics of these objects.

Monitoring Orbits: Calculating Future Trajectories

Once an NEO is discovered, the next step is to calculate its orbit and predict its future trajectory. This requires multiple observations of the object over time to determine its position, speed, and direction. The LSST's frequent imaging cadence is ideal for this task, as it will provide repeated observations of each NEO, allowing astronomers to track its motion across the sky.

By measuring the positions of NEOs in multiple images taken over several nights, the LSST will enable scientists to calculate accurate orbital parameters. These parameters can then be used to predict the NEO's future path and assess whether it poses a threat to Earth. If a potential impact is identified, scientists can issue early warnings, providing time to develop mitigation strategies.

The LSST's ability to track NEOs over long periods will also help improve our understanding of how these objects' orbits change over time. NEOs are subject to various forces that can alter their trajectories, including gravitational interactions with planets, collisions with other objects, and the Yarkovsky effect—a subtle force caused by the emission of heat from an asteroid's surface. By monitoring NEOs over the course of its 10-year survey, the LSST will provide valuable data on how these forces affect their orbits, improving our ability to predict future movements.

Planetary Defense: Mitigating Impact Risks

One of the primary goals of NEO detection is to mitigate the risk of an impact with Earth. While the chances of a large-scale impact are low, the potential consequences are severe enough to warrant ongoing monitoring and preparation. The LSST will play a critical role in planetary defense by providing early warnings of potential impacts, giving scientists and policymakers the time needed to respond.

92

If an NEO is identified as a potential impactor, there are several possible mitigation strategies, depending on the size and trajectory of the object. These strategies include:

1. **Deflection**: One of the most widely discussed methods of preventing an impact is to deflect the NEO's path by altering its trajectory. This could be done using a spacecraft to either push or pull the object, or by using a kinetic impactor to physically strike the NEO and change its orbit. The earlier an NEO is detected, the more time there is to implement a deflection strategy.

2. **Disruption**: In some cases, it may be necessary to break the NEO into smaller pieces that would burn up in Earth's atmosphere. This could be achieved using a nuclear device or other means of disruption. However, this strategy carries significant risks, as the fragments could still cause damage if they reach the surface.

3. **Evacuation and Preparation**: For smaller NEOs or objects that are detected too late for deflection or disruption, the best option may be to evacuate the area of the projected impact and prepare for the event. Early detection and accurate predictions of the

93

impact location are critical for ensuring that appropriate measures can be taken.

The LSST's ability to detect NEOs early and track their orbits with precision will provide the information needed to implement these mitigation strategies. By providing accurate and timely data, the LSST will contribute to the global effort to protect Earth from the threat of NEO impacts.

Collaborating with Global Planetary Defense Networks

The LSST's contributions to NEO detection and planetary defense will not occur in isolation. The observatory will work in conjunction with other space agencies and organizations that are dedicated to monitoring and responding to NEO threats. These include NASA's *Planetary Defense Coordination Office* (PDCO), the European Space Agency's (ESA) *Space Situational Awareness* (SSA) program, and the International Asteroid Warning Network (IAWN).

NASA's PDCO is responsible for detecting, tracking, and characterizing potentially hazardous asteroids and comets, as well as coordinating efforts to mitigate impact threats. The LSST's data will be shared with NASA and other agencies, providing valuable information to support their planetary defense efforts.

In addition, the LSST will contribute to the work of the IAWN, a global network of observatories and space agencies that share data on NEOs. The IAWN serves as a clearinghouse for information on potential impact risks, and the LSST's contributions will help ensure that the latest data is available to the global scientific community.

The Future of NEO Detection with the LSST

As the LSST prepares to begin its 10-year survey, its role in NEO detection and planetary defense is set to revolutionize our ability to monitor and respond to potential threats from space. By providing a comprehensive, real-time view of the sky, the LSST will enhance our understanding of NEO populations, improve orbital predictions, and enable early detection of potential impactors.

The data collected by the LSST will not only help protect Earth from the threat of asteroid impacts but also contribute to our broader understanding of the solar system. By studying the orbits, compositions, and behaviors of NEOs from the survey will revolutionize our understanding of how these small bodies move, interact with planetary bodies, and evolve over time. This information will also deepen our understanding of the risks these objects pose to Earth and inform the strategies used to mitigate those risks.

The Vera C. Rubin Observatory's LSST Camera will be a game-changer for planetary defense. Its ability to detect and track Near-Earth Objects, both known and previously undiscovered, will significantly enhance our ability to monitor and mitigate potential threats to Earth. By providing real-time data on the positions, trajectories, and behaviors of NEOs, the LSST will contribute to global efforts to protect our planet from the dangers of asteroid impacts. Its role in planetary defense, in conjunction with international organizations and space agencies, marks a significant step forward in securing the safety of our solar system for future generations.

CHAPTER 8

A New Era for Astronomy: Scientific and Public Engagement

The Vera C. Rubin Observatory and its 3.2-gigapixel digital camera, as part of the *Legacy Survey of Space and Time* (LSST), represent a transformative step in modern astronomy. Over the course of a decade, the LSST will systematically map the southern sky, collecting an unprecedented volume of data that will revolutionize our understanding of the universe. However, the observatory's significance goes beyond just the scientific community. It has the potential to engage the public, inspire new generations of astronomers, and foster a deeper appreciation of the cosmos among people worldwide.

In this final chapter, we will explore the broader implications of the LSST for both the scientific community and the public. We will examine how its discoveries could reshape our understanding of the universe and discuss the ways in which the project will inspire future generations and increase public engagement with astronomy.

Redefining Astronomy: LSST's Scientific Impact

The scientific implications of the Vera C. Rubin Observatory are vast, covering almost every major area of astrophysics and cosmology. The LSST's combination of deep, wide-field imaging, and repeated sky coverage over 10 years will produce the most detailed and comprehensive map of the southern sky ever created. This data will be used to address some of the most profound questions in science, including the nature of dark matter and dark energy, the formation and evolution of galaxies, the behavior of variable stars, and the potential for hazardous Near-Earth Objects (NEOs).

Dark Matter and Dark Energy: The Universe's Greatest Mysteries

One of the most significant contributions the LSST will make is in the study of dark matter and dark energy, which together make up approximately 95% of the universe. These components remain largely mysterious, but their gravitational effects shape the large-scale structure of the cosmos. Dark energy, in particular, is responsible for the accelerating expansion of the universe.

The LSST will help map the distribution of dark matter through techniques like gravitational lensing, where light from distant galaxies is bent by the gravity of intervening dark matter. By observing the distortion of background light as it passes through regions of dark matter, the LSST will create detailed maps of where dark matter is located and how it affects the motion of galaxies.

Additionally, the LSST will provide data to better understand dark energy by observing distant Type Ia supernovae— exploding stars that act as "standard candles" for measuring cosmic distances. By studying the redshift of these supernovae, astronomers will refine their understanding of the universe's expansion rate and gain new insights into the role dark energy plays in shaping the cosmos.

These discoveries have the potential to radically alter our understanding of the universe's underlying structure. They may confirm existing theories or point toward new physics that challenges the standard model of cosmology.

Charting the Evolution of Galaxies

The LSST will capture images of billions of galaxies across a wide range of distances and epochs, providing a detailed record of how galaxies form, grow, and evolve over cosmic time. This data will allow astronomers to trace the growth of galaxies from the early universe to the present day, shedding light on the processes that govern galaxy formation, including mergers, interactions, and star formation.

One key area of research will involve the LSST's ability to observe galaxy clusters, the largest gravitationally bound structures in the universe. These clusters contain thousands of galaxies and are rich sources of information about dark matter and the large-scale structure of the universe. By studying the motions and distribution of galaxies in these clusters, the LSST will help scientists better understand how galaxies form within the cosmic web of dark matter.

The observatory's wide-field imaging capabilities will also be instrumental in identifying new types of galaxies, including faint or distant galaxies that have eluded previous surveys. By cataloging such a large number of galaxies, the LSST will offer a unique opportunity to study the statistical properties of these systems, providing crucial data for testing theories of galaxy evolution.

Planetary Defense and Near-Earth Objects

Another major scientific benefit of the LSST will be its contribution to planetary defense. As discussed in earlier chapters, the observatory's ability to detect and track NEOs is a critical tool in protecting Earth from potential asteroid impacts. By providing early detection and accurate tracking of these objects, the LSST will enhance global efforts to mitigate the risks posed by potentially hazardous asteroids.

The data collected by the LSST will also improve our understanding of the small-body populations in the solar system, including asteroids and comets. These objects are remnants of the early solar system, and studying them can provide insights into the conditions that prevailed during the formation of the planets.

In addition to improving our knowledge of NEOs, the LSST will contribute to the discovery and tracking of other solar system objects, including distant trans-Neptunian objects (TNOs) in the outer solar system. These objects hold clues to the structure and evolution of the solar system, and their discovery will open new frontiers in planetary science.

A New Generation of Astronomers: Inspiring Public Engagement

Beyond its scientific contributions, the Vera C. Rubin Observatory has the potential to inspire a new generation of astronomers and foster public engagement with science. In an era where science and technology play an increasingly important role in society, projects like the LSST can ignite curiosity and a sense of wonder about the universe, encouraging people from all walks of life to explore astronomy.

Democratizing Data: Open Access and Citizen Science

One of the key features of the LSST is its commitment to open access. The vast majority of the data collected by the observatory will be made freely available to the public, allowing not just professional astronomers, but also students, educators, and citizen scientists to explore the universe alongside experts.

The open-access model is a significant departure from traditional astronomical surveys, where data is often restricted to specific research teams or institutions. By making its data available to everyone, the LSST will democratize astronomy, giving people around the world the opportunity to contribute to scientific discoveries.

Citizen science projects, such as those hosted by *Zooniverse*—a platform that allows the public to participate in real scientific research—will likely play a key role in the analysis of LSST data. With billions of objects to catalog and countless phenomena to discover, there will be no shortage of opportunities for citizen scientists to get involved. This

102

engagement not only supports scientific research but also fosters a deeper connection between the public and the scientific process.

Inspiring the Next Generation: Educational Outreach

The LSST's discoveries and data will provide rich material for educational outreach programs, both in classrooms and through online platforms. The observatory's website, social media presence, and educational initiatives will ensure that the public has access to the latest discoveries, complete with interactive tools to explore the data.

For teachers and students, the LSST's open-access data offers a wealth of resources for engaging with real scientific research. Students will have the opportunity to work with the same data used by professional astronomers, offering a hands-on experience that can inspire interest in STEM (science, technology, engineering, and mathematics) fields.

In addition, the LSST's ability to capture transient events, such as supernovae and gamma-ray bursts, will provide exciting opportunities for educational outreach. By showcasing the dynamic nature of the universe, these events can spark curiosity and encourage students to pursue careers in astronomy or related fields.

Public Engagement with Astronomy

The LSST has the potential to bring astronomy into the public consciousness in new and exciting ways. The visual appeal of the observatory's high-resolution images, coupled with its real-time observations of transient events, will capture the public's imagination and inspire awe at the scale and beauty of the universe.

Public interest in space exploration has grown in recent years, thanks in part to high-profile missions like NASA's *Perseverance* rover on Mars, as well as the increasing involvement of private companies like SpaceX. The LSST will build on this momentum by offering the public a unique window into the cosmos, providing a steady stream of new discoveries and insights into the nature of the universe.

As the observatory captures stunning images of galaxies, nebulae, and other celestial objects, these images will be shared widely across the internet and social media platforms. This visual content has the potential to reach millions of people, sparking interest in astronomy and encouraging people to learn more about the universe.

The Legacy of the Vera C. Rubin Observatory

The Vera C. Rubin Observatory is poised to leave a lasting legacy in the field of astronomy, not only through its scientific contributions but also by inspiring the public and engaging new generations of astronomers. The observatory's ability to capture a dynamic view of the universe, combined with its commitment to open access and public engagement, sets it apart from previous astronomical projects.

Over the next decade, the LSST will generate a wealth of data that will transform our understanding of the cosmos, offering new insights into everything from dark matter and dark energy to the behavior of variable stars and the threat of NEOs. These discoveries will reshape our understanding of the universe and push the boundaries of what we know about space.

At the same time, the LSST's open-access model will allow the public to participate in this journey of discovery, fostering a sense of connection to the cosmos and encouraging people from all backgrounds to engage with astronomy. As we look to the future, the Vera C. Rubin Observatory stands as a testament to the power of collaboration, innovation, and the pursuit of knowledge—

values that will continue to inspire both the scientific community and the public for generations to come.

The Vera C. Rubin Observatory's digital camera and the LSST are ushering in a new era for astronomy. With its ability to capture vast amounts of data, inspire citizen scientists, and educate the public, this observatory represents a turning point in our relationship with the cosmos. Whether it's uncovering the mysteries of dark matter, protecting Earth from asteroids, or inspiring the next generation of discoverers, the LSST will play a vital role in bringing the wonders of the universe to the world.

CONCLUSION

The Game-Changing Telescope Shaping Astronomy's Future

As we draw to the end of this exploration into the profound impact of the world's largest digital camera and the game-changing telescope it resides in, it's essential to reflect on the incredible strides astronomy has made—and will continue to make—because of these advancements. The Vera C. Rubin Observatory, equipped with the 3.2-gigapixel Legacy Survey of Space and Time (LSST) Camera, represents a leap forward in both technology and our understanding of the cosmos. This telescope, with its digital camera, is set to revolutionize not only how we explore the universe but also how we interpret the very structure of the cosmos and engage with the field of astronomy.

A Revolutionary Tool for Cosmic Exploration

The largest digital camera ever constructed, mounted on the Rubin Observatory, is not just a technological marvel but also a symbol of what's possible when human ingenuity is applied to the mysteries of the universe. This camera is designed to capture 15 terabytes of data each night over a

10-year period, providing a window into the universe's dynamic events, from the slow evolution of galaxies to the fleeting appearance of supernovae. This vast dataset will allow scientists to construct a detailed, time-lapse movie of the universe, tracking changes that were previously beyond our observational capabilities.

The construction and deployment of such a digital camera signal an evolution in astronomical observation. By covering an area of the sky larger than any previous survey and repeating its observations regularly, the LSST Camera will detect not only distant and faint objects but also those that change in real-time, such as asteroids, variable stars, and transient events like gamma-ray bursts. This ability to monitor the sky for change is truly revolutionary, providing a dynamic view of the universe and uncovering events and objects that might otherwise go unnoticed.

Expanding the Frontiers of Astrophysics

The data gathered by the LSST will serve as the foundation for groundbreaking research in several fields of astrophysics. From dark matter to dark energy, the LSST Camera will provide the kind of precision and depth of observation necessary to map the universe's invisible scaffolding. By capturing the faint distortions in light caused by gravitational

108

lensing—an effect where light from distant galaxies is bent by dark matter—the LSST will offer scientists critical insights into how dark matter is distributed throughout the cosmos.

Likewise, the role of dark energy in the accelerating expansion of the universe will be studied in unprecedented detail. The LSST's ability to track the movement of galaxies and measure their redshift will allow scientists to refine their models of cosmic expansion. These findings could lead to revolutionary theories about the fate of the universe and redefine our understanding of the underlying forces that shape the cosmos.

In addition to these large-scale phenomena, the LSST will shed light on smaller, more immediate celestial objects. Near-Earth Objects (NEOs), such as asteroids and comets, will be tracked with unparalleled precision. The LSST will be able to detect these objects, calculate their orbits, and assess the potential risk they pose to Earth. This capability will make significant contributions to planetary defense, providing early warnings of potentially hazardous asteroids and giving humanity time to devise mitigation strategies.

The Impact on Scientific Collaboration

The scale of data generated by the LSST—expected to reach 15 petabytes over the duration of its survey—presents challenges that will push the boundaries of data science. Processing, storing, and analyzing this vast quantity of information will require state-of-the-art computational techniques, including artificial intelligence and machine learning. As a result, the LSST project is a harbinger of a new era of "big data" astronomy, where scientific breakthroughs will depend on our ability to manage and interpret massive datasets.

However, the technological innovations spurred by the LSST are not limited to data management alone. The observatory's open-access policy, which makes the data available to scientists worldwide, will democratize astronomical research. For the first time, astronomers, data scientists, and even citizen scientists from across the globe will have access to the same wealth of information, fostering collaboration and innovation across borders. By breaking down the traditional barriers to data access, the Rubin Observatory is setting a new standard for openness and inclusivity in scientific research.

110

Engaging the Public and Inspiring the Next Generation

One of the most exciting aspects of the Vera C. Rubin Observatory's mission is its potential to inspire the public and engage a new generation of scientists, engineers, and space enthusiasts. Astronomy has always had a unique ability to capture the imagination, and the breathtaking images that will be produced by the LSST Camera will undoubtedly stir curiosity and wonder in people of all ages.

The LSST project will provide educational opportunities for students, offering them the chance to work with real scientific data. As citizen science platforms continue to grow, the general public will also have the chance to participate in this cosmic exploration. Platforms like *Zooniverse*, which has previously engaged millions of people in scientific discoveries, will likely see an influx of participants eager to sift through the LSST's vast datasets, potentially uncovering new phenomena and aiding researchers in significant ways.

In addition to its scientific value, the LSST's ability to engage the public represents a crucial investment in the future of science. As governments and private institutions increasingly turn their attention toward space exploration and technological innovation, cultivating public interest in

111

these fields is vital. By inspiring students, amateurs, and enthusiasts to take part in scientific discovery, the LSST helps ensure that the next generation of astronomers and space pioneers will be ready to tackle the challenges of tomorrow.

Paving the Way for Future Observatories

The success of the Vera C. Rubin Observatory and its LSST Camera will undoubtedly influence the design and development of future telescopes. Already, the LSST has set a new standard for sky surveys, combining a wide field of view with high-resolution imaging and rapid data collection. Future observatories may build on these advancements, incorporating similar technologies to achieve even greater feats of observation.

Moreover, the collaborative model pioneered by the LSST—where data is shared openly with scientists worldwide—could become the norm for future large-scale scientific projects. By emphasizing cooperation and inclusivity, the LSST has shown that scientific discovery can be accelerated when researchers pool their resources and expertise.

As new observatories come online in the coming decades, they will likely adopt and improve upon the techniques used by the LSST. Projects like the European Southern Observatory's Extremely Large Telescope (ELT) and NASA's James Webb Space Telescope (JWST) will benefit from the groundwork laid by the LSST, particularly in the areas of big data management, AI-driven analysis, and public engagement.

Shaping the Future of Astronomy

As we conclude our journey through the Vera C. Rubin Observatory and its game-changing telescope, it's clear that the LSST represents a pivotal moment in the history of astronomy. The combination of cutting-edge technology, a vast scientific scope, and an unprecedented commitment to open access makes this project truly revolutionary. By capturing a time-lapse view of the universe, the LSST will provide insights that deepen our understanding of the cosmos and uncover mysteries that we have yet to even imagine.

The future of astronomy is bright, and the LSST is leading the way. Its discoveries will help answer fundamental questions about the nature of the universe while also raising new ones, pushing the boundaries of what we know and what we can achieve. From dark matter to planetary defense, from the formation of galaxies to the search for potentially habitable worlds, the LSST Camera will help unlock the secrets of the universe and set the stage for future explorations.

The Vera C. Rubin Observatory is not just a telescope; it's a gateway to the future. With its digital camera, the largest ever built, it will shape the future of astronomy and inspire generations to look up, wonder, and explore the cosmos. This is only the beginning of a new era—an era where the boundaries of human knowledge will be expanded, and where the universe's most profound secrets will slowly come to light.

www.ingramcontent.com/pod-product-compliance
Lightning Source LLC
Chambersburg PA
CBHW071523220526
45472CB00003B/1130